# LEONARD BASKIN

HEINEMANN · LONDON

Giant Moa / *Diornis maximus* / EXTINCT C. 17TH CENTURY

EXTINCT
CREATVRES

# CHRISTMAS ISLAND MUSK SHREW

*Crocidura fulginosa trichura*

EXTINCT C. 1900

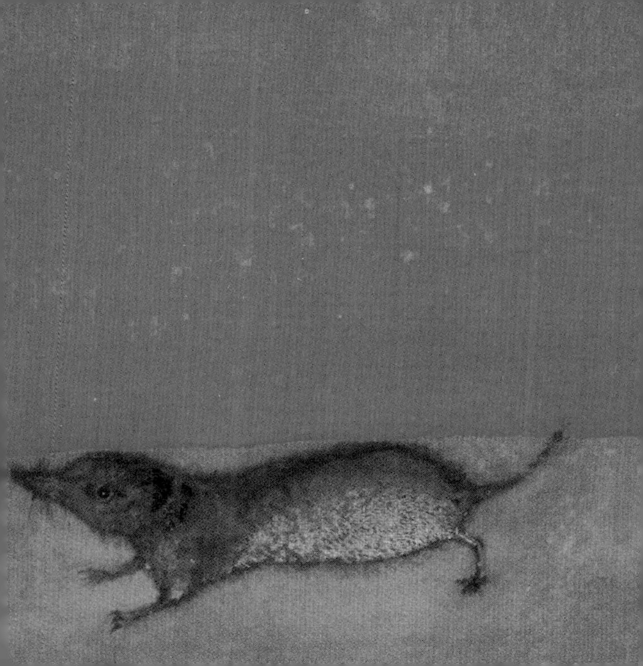

PYGMY HOG

*Sus slavanius*

ENDANGERED. THIRTY TO FORTY SEEN IN 1977

## PTERODACTYL

*Pterodactylus*

EXTINCT C. 100,000,000 YEARS AGO

# STEPHEN ISLAND WREN

*Xenicus lyalli*

EXTINCT C. 1894

## MEXICAN GRIZZLY BEAR

*Ursus horribilis nelsoni*

ENDANGERED. LAST SEEN IN EARLY 1960S

# STEGOSAURUS

*Stegosaurus*

EXTINCT C. 150,000,000 YEARS AGO

FLORIDA BLACK WOLF

*Canis rufus floridanus*

ENDANGERED. LAST SEEN IN 1917

DODO

*Raphus cucullatus*

EXTINCT C. 1860

SABRE-TOOTH TIGER

*Smilodon californicus*

EXTINCT C. 50,000 YEARS AGO

HORNED DINOSAUR

*Triceratops*

EXTINCT C. 65,000,000 YEARS AGO

## MIYAKO KINGFISHER

*Halcyon miyakoensis*

EXTINCT C. 1887

William Heinemann Ltd, 10 Upper Grosvenor Street,
London W1X 9PA
LONDON · MELBOURNE · TORONTO
JOHANNESBURG · AUCKLAND
First Published in Great Britain 1983


Manufactured in Singapore.
Published by arrangement with Pantheon Books,
a division of Random House, Inc.
434 04907 7

Leonard Baskin's watercolour miniatures of the Giant Moa, Stephen Island Wren, Florida Black Wolf and Miyako Kingfisher are based on illustrations from *The Doomsday Book of Animals: A Natural History of Vanished Species* by David Day. Copyright © 1981 by London Editions/David Day. Used by permission of Ebury Press.

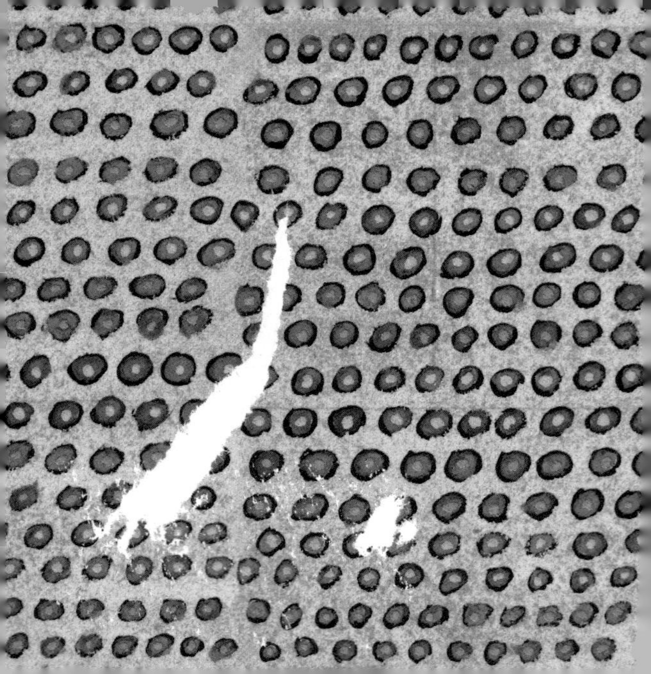

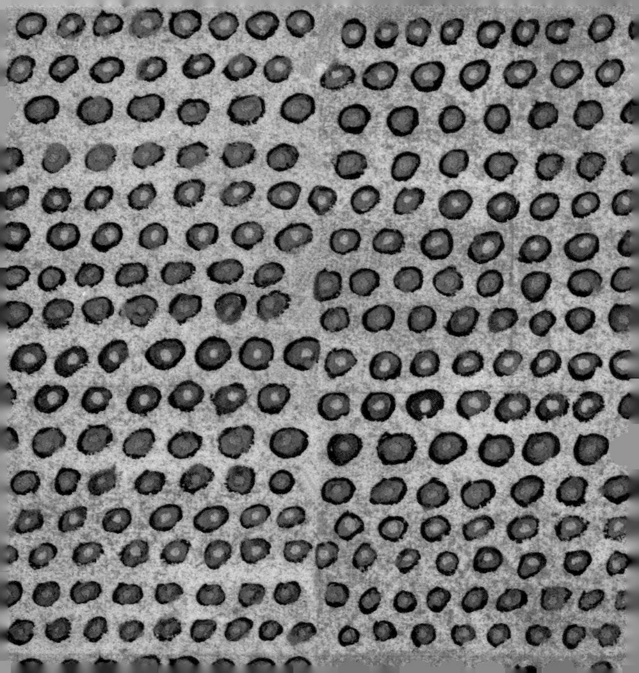

Mosquito / *Culex pipiens*

# LEONARD BASKIN

HEINEMANN · LONDON

# INSECTS

BUMBLEBEE

*Bombus lapidarius*

## STAG BEETLE

*Lucanus cervus*

## DEATH'S-HEAD HAWK MOTH

*Acherontia atropos*

# GIANT WOODWASP

*Urocerus gigas*

DRAGONFLY

*Aeschna cyanea*

KATYDID

*Microcentium rhombifolium*

## DIVING BEETLE

*Dytiscus latissimus*

GRASSHOPPER

*Omocestus viridulus*

## SPITTLE BUG

*Cercopis vulnerata*

# PRAYING MANTIS

*Mantis religiosa*

## TIGER MOTH

*Arctia caja*

William Heinemann Ltd, 10 Upper Grosvenor Street,
London W1X 9PA
LONDON · MELBOURNE · TORONTO
JOHANNESBURG · AUCKLAND
First Published in Great Britain 1983


Manufactured in Singapore.
Published by arrangement with Pantheon Books,
a division of Random House, Inc.
434 04907 7

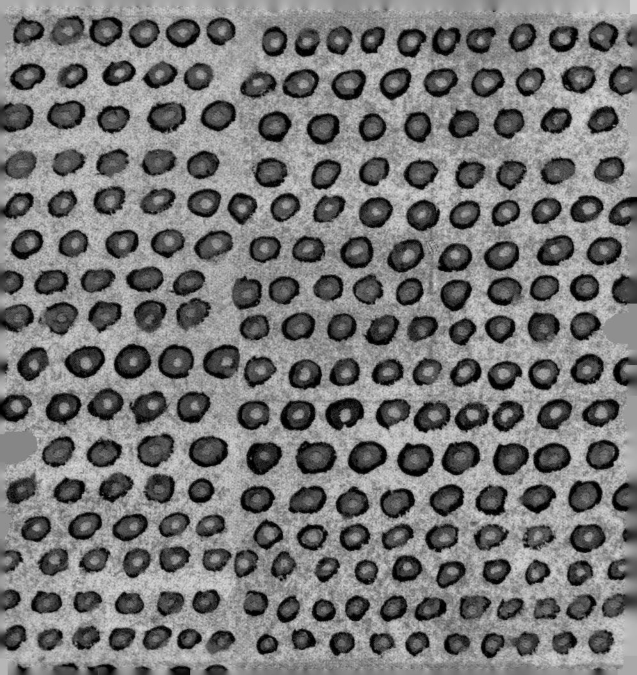

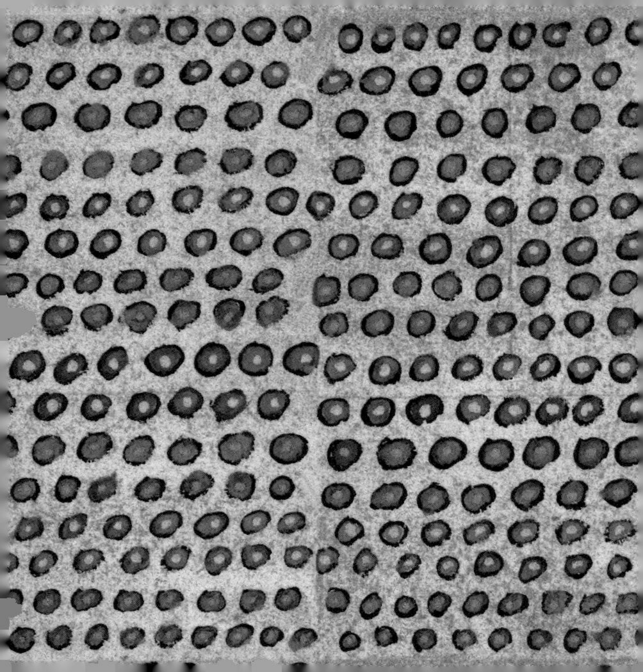

# LEONARD BASKIN

HEINEMANN · LONDON

Canary / *Serinus canarius*

DOMESTIC
ANIMALS

GOAT

*Capra hireus*

# HOUSE MOUSE

*Mus musculus*

DOG

*Canis familiaris*

PIG

*Sus scrofa*

DORSET DOWN RAM

*Ovis aries*

# GOOSE

*Anser anser*

CHAROLAIS BULL

*Bos taurus*

COCKEREL

*Gallus gallus*

## CAT

*Felis domestica*

# CHICK

*Gallus gallus*

# SHEEP

*Ovis aries*

William Heinemann Ltd, 10 Upper Grosvenor Street,
London W1X 9PA
LONDON · MELBOURNE · TORONTO
JOHANNESBURG · AUCKLAND
First Published in Great Britain 1983


Manufactured in Singapore.
Published by arrangement with Pantheon Books,
a division of Random House, Inc.
434 04907 7

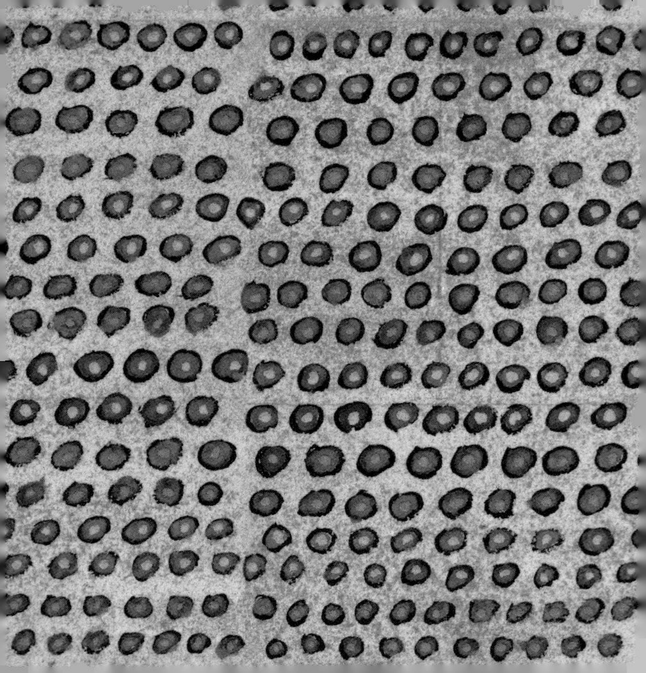

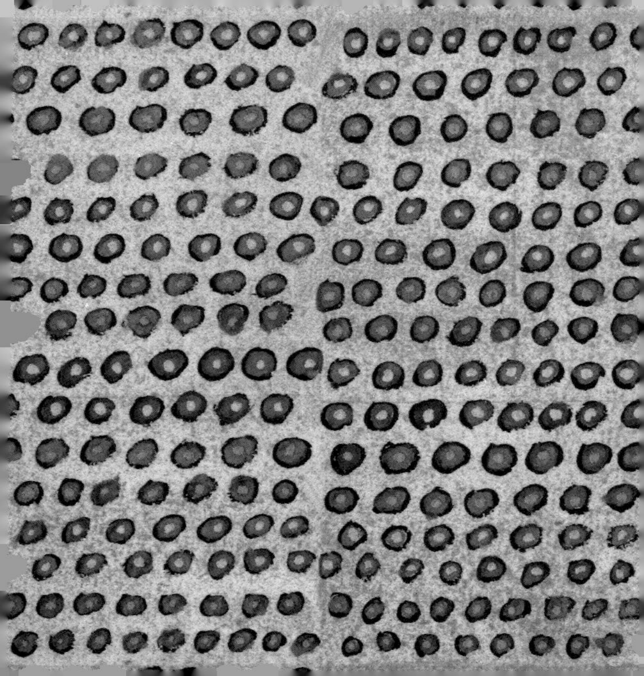

# LEONARD BASKIN

HEINEMANN · LONDON

Butterfly Fish / *Chelmon rostratus*

SEVEN FISH
ONE CRAB
ONE LOBSTER
&
ONE OCTOPUS
& ONE PRAWN

## ATLANTIC MACKEREL

*Scomber scomber*

## PIKE

*Esox lucius*

## OCTOPUS

*Octopus dofleini*

## GREEN CRAB

*Carcinus maenas*

# PLAICE

*Pleuronectes platessa*

## DEEP SEA PRAWN

*Sergestes*

TRIGGERFISH

*Balistapus undulatus*

## LOBSTER

*Homarus gammarus*

SNIPEFISH

*Macrorhamphosus elevatus*

STINGRAY

*Taeniura lymma*

SKATE

*Raja texana*

William Heinemann Ltd, 10 Upper Grosvenor Street,
London W1X 9PA
LONDON · MELBOURNE · TORONTO
JOHANNESBURG · AUCKLAND
First Published in Great Britain 1983


Manufactured in Singapore.
Published by arrangement with Pantheon Books,
a division of Random House, Inc.
434 04907 7

B-ue
[illegible] (R)
March 1985